W9-CEG-222

THE ELEMENTS

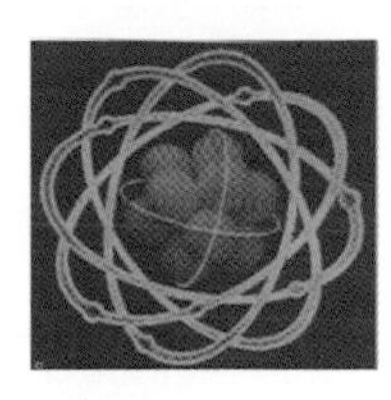

Fluorine

Tom Jackson

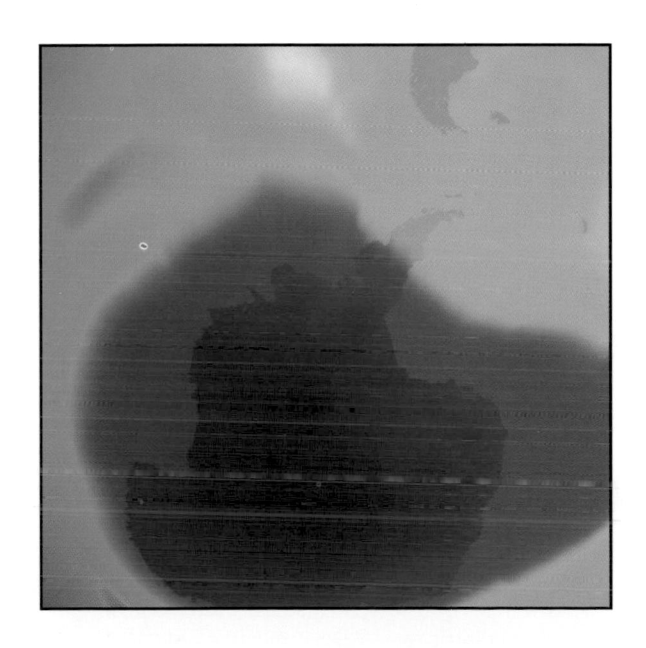

BENCHMARK BOOKS
MARSHALL CAVENDISH
NEW YORK

Benchmark Books
Marshall Cavendish
99 White Plains Road
Tarrytown, New York 10591

www.marshallcavendish.com

Library of Congress Cataloging-in-Publication Data

Jackson, Tom, 1972–
Fluorine / Tom Jackson.
p. cm. — (The elements)
Includes index.
Contents: What is fluorine? — Fluorine in nature —
Special characteristics — The search for fluorine — Chemistry and compounds —
Fluorine and manufacturing — Fluorine and fuel — Fluorine and health —
Fluorocarbons — Periodic table — Chemical reactions.
ISBN 0-7614-1549-1
1. Fluorine—Juvenile literature. [1. Fluorine.] I. Title. II.
Elements (Benchmark Books)
QD181.F1J33 2004
546'.731—dc21

2003043841

Printed in China

Picture credits

Front Cover: Air Products & Chemicals Ltd.
Back Cover: Bethlehem Steel

Air Products & Chemicals Ltd.: 8, 14, 17
Bethlehem Steel: 16
Corbis: Bettmann 22, C/B Productions 4, Jose Manuel Sanchis Calvete *iii,* 7 (*top*),
William Gottlieb 21, Philip Gould 25, John Heseltine 6
Dupont: 26
Gift of Mlle. Florence Marinot: Collection of The Corning Museum of Glass,
Corning, New York 18
NASA: *i,* 20, 24
The Nobel Foundation: 10
Dr. Graham Sandfold/Durham University: 12
Science & Society Picture Library: Science Museum 11
Science Photo Library: 9, 27, Herve Berthoule 30, Mark A. Schneider 7 (*bottom*),
Sheila Terry 23, U.S. Department of Energy 19

Series created by The Brown Reference Group plc
Designed by Sarah Williams
www.brownreference.com

Contents

What is fluorine?	4
Fluorine in nature	6
Special characteristics	8
The search for fluorine	9
Chemistry and compounds	12
Fluorine and manufacturing	16
Fluorine and fuel	19
Fluorine and health	21
Fluorocarbons	24
Periodic table	28
Chemical reactions	30
Glossary	31
Index	32

What is fluorine?

Fluorine belongs to a group of elements called the halogens. Other halogens include bromine, chlorine, and iodine. All the halogens are highly reactive, but fluorine is the most reactive of all the elements. Fluorine combines with most of the other elements to form compounds called fluorides.

Fluorine is a pale yellow gas as an element, and it is also very dangerous. Breathing in just a tiny amount of fluorine gas damages the lungs. But fluorine is not all bad. In fact, fluorine's reactivity makes it a very useful element. Perhaps the most common use of a fluorine compound is in toothpaste. The fluorine compounds are incorporated into tooth enamel, which makes our teeth stronger. Other fluorine compounds are used to isolate aluminum from its ores, remove radioactive uranium atoms from nuclear reactors, and make the nonstick surfaces of frying pans.

The fluorine compounds in toothpaste strengthen our teeth so they can resist attack from harmful bacteria.

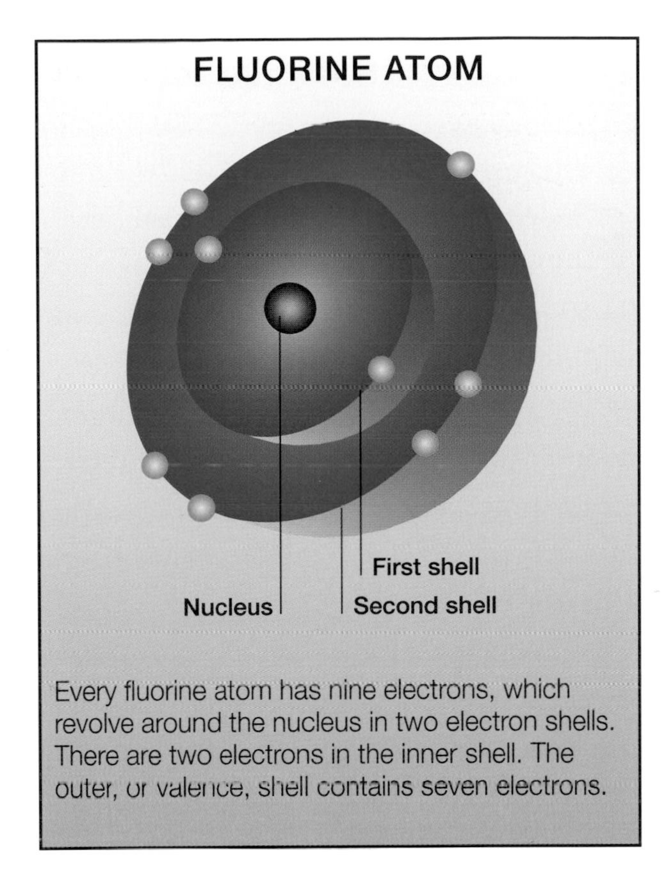

Every fluorine atom has nine electrons, which revolve around the nucleus in two electron shells. There are two electrons in the inner shell. The outer, or valence, shell contains seven electrons.

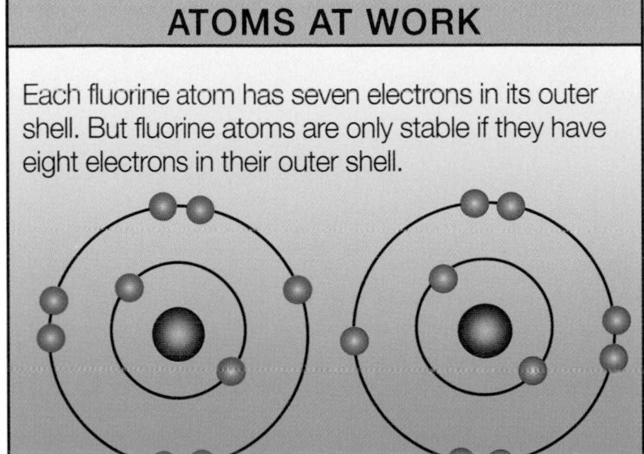

Each fluorine atom has seven electrons in its outer shell. But fluorine atoms are only stable if they have eight electrons in their outer shell.

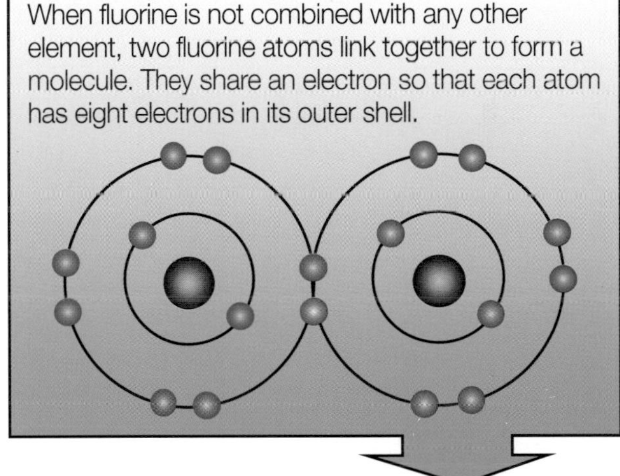

When fluorine is not combined with any other element, two fluorine atoms link together to form a molecule. They share an electron so that each atom has eight electrons in its outer shell.

Fluorine gas is made up of these pairs of atoms, which are called diatomic molecules. The formula of diatomic fluorine is written like this:

F_2

Inside the atom

Atoms are the building blocks of the elements. Atoms consist of tiny particles called protons, neutrons, and electrons. The protons and neutrons cluster in the dense nucleus at the center of each atom. The number of protons is given by the atomic number. Fluorine has an atomic number of nine, so there are nine protons in the nucleus of each atom. The protons and neutrons combine to give the atom its mass. Fluorine has an atomic mass of 19, which means each atom has ten neutrons.

Electrons revolve around the nucleus in layers called electron shells. Fluorine atoms contain nine electrons that revolve around the nucleus in two shells. The inner shell contains two electrons, and the outer shell has seven electrons. The outer shell is incomplete because it actually has room for eight electrons. That is why fluorine is so reactive. In a chemical reaction it gains an extra electron to fill its outer shell.

Fluorine in nature

Fluorine does not exist as an element in nature, but its compounds make up 0.065 percent of Earth's crust. The amount of fluorine on Earth is tiny compared to elements such as aluminum, carbon, iron, and oxygen, but it is more common than all the other halogens except chlorine.

Forms of fluorine

The number of protons and electrons in an atom of a specific element is always the same. But sometimes the atoms of some elements have more or fewer neutrons than others. Those atoms are called isotopes. Unlike most other elements, fluorine has only one natural isotope, called fluorine-19. The number 19 stands for the number of protons plus the number of neutrons in the nucleus.

Mineral content

Most of the fluorine in Earth's crust is found as crystalline compounds called minerals. Fluorspar (fluorite) is the most common fluorine-containing mineral. Fluorspar consists mainly of calcium fluoride (CaF_2), but about one-fifth

Fluorspar deposits in the Blue John Cavern near Castleton, Britain. Most fluorine in nature is found combined with calcium as the mineral fluorspar.

DID YOU KNOW?

TOPAZ

Topaz is a common gemstone that has been used to decorate jewelry for centuries. This shiny mineral consists of aluminum, silicon, hydrogen, oxygen, and fluorine. Naturally occurring topaz is golden brown or yellow, and it may be cloudy or transparent. A few natural deposits consist of a highly prized blue topaz. Natural deposits are rare, however, so most blue topaz is made artificially.

Topaz is one of the hardest minerals in nature. A single crystal can weigh more than one ton. The best places to find topaz include Brazil, Pakistan, Mexico, Russia, California, and Utah.

of the mineral contains the elements yttrium and cerium. Fluorspar is a glassy crystal. Large parts of China, southern Africa, France, Mexico, and Russia have deposits of fluorspar.

Fluorspar has been used to refine metals for thousands of years. When fluorspar is heated with a metal until it turns into a molten liquid, any impurities "flow" away from the metal and react with fluorine in the fluorspar. The mineral is named for this function—*fluere* is the Latin word for "flow." When fluorspar was found to be a source of fluorine, the element was given a similar name.

Another important fluorine mineral is cryolite (Na_3AlF_6). Greenland is a major source of cryolite, but this mineral is rare elsewhere. Most of the fluorine used in industry is manufactured from fluorspar.

Phosphate rock

Other natural sources of fluorine are minerals called apatites, which consist of calcium, phosphorus, and tiny amounts of fluorine. Apatite deposits are common in the United States, Russia, and North Africa. Since fluorine makes up just a small part of these minerals, apatites are far more valuable for the phosphorus they contain. Phosphorus is important as a fertilizer for crops.

Crystals of the minerals fluorspar (shown in blue) and quartz (red) fluoresce or glow if they are placed under an ultraviolet lamp.

Special characteristics

At standard room temperature and pressure, fluorine is a pale yellow gas. Fluorine gas is slightly heavier than air, so it gradually sinks to the ground if it is released. If fluorine is cooled to a chilly –307 °F (–188 °C), the gas turns into a yellow liquid. If this liquid is cooled further to –363 °F (–219 °C), it freezes into an almost colorless crystal.

Fluorine is lighter and has a lower melting point and freezing point than all the other halogens. At standard room temperature and pressure, chlorine is a heavy, green gas; bromine is a brown liquid; and iodine is a purple solid.

A fluorine laboratory in Amagasaki near Osaka in Japan. Fluorine gas is so dangerous that it can only be used in carefully controlled conditions.

FLUORINE FACTS	
Atomic number:	9
Atomic mass:	18.9984032
Melting point:	–363.32 °F (–219.62 °C)
Boiling point:	–306.62 °F (–188.12 °C)
Density:	0.001696 grams per cubic centimeter
Solid/liquid/gas:	Gas at room temperature
Isotopes:	One naturally occurring isotope—fluorine-19
Name's origin:	The word *fluorine* comes from the Latin *fluere,* which means "flow"

A reactive element

Fluorine gas has a sharp odor. Only chemists and engineers use pure fluorine because it is very expensive to make and dangerous to handle. If someone breathes in fluorine gas by accident, it burns their skin and lungs. This is because the fluorine reacts with the water in the body's cells to produce a corrosive chemical called hydrofluoric acid (HF).

Fluorine's reactivity is caused by the arrangement of electrons in its outer shell. A fluorine atom has seven electrons in its outer shell. It needs eight electrons to be stable. During a reaction, fluorine atoms steal electrons from other atoms to create full outer shells. In this way, fluorine forms very stable compounds. Once fluorine's outer shell is full, the atom will not give up the electron very easily to form the reactive gas again.

The search for fluorine

Fluorspar is the most common mineral containing fluorine. For thousands of years, people have used fluorspar to produce pure metals such as iron. The first person to describe the mineral in detail was German mineralogist and physician Georg Bauer (1494–1555), who is known by his Latin name Agricola.

An engraving of German mineralogist Agricola. In 1546, Agricola wrote a book called De Natura Fossilium *("The Nature of Fossils"), in which he described the mineral fluorspar.*

DISCOVERERS

DANGEROUS PURSUIT

Many chemists tried to isolate fluorine using electrolysis before Moissan's successful attempt in 1886. Most passed electricity through a solution of hydrogen fluoride, but many chemists were injured or even killed during the procedure.

Like fluorine, hydrogen fluoride is an extremely corrosive gas and burns the lungs and skin. Belgian chemist Paulin Louyet died from inhaling hydrogen fluoride vapors. Passing electricity through a hydrogen fluoride solution produces fluorine and hydrogen gas at each electrode. If these two gases then mix, they recombine into hydrogen fluoride in an explosive reaction. In 1869, English chemist George Gore made a small amount of fluorine in this way, but he did not anticipate the explosive consequences. Many other chemists were forced to give up their search for fluorine for the sake of their health.

Moissan overcame all these difficulties, but he also died at a young age—less than a year after being honored for his work. Maybe the fluorine took its toll on his health, too?

By 1771, chemists had learned to make an acidic gas called hydrogen fluoride (HF) by heating fluorspar with a strong acid. It was dangerous work for these early chemists. Hydrogen fluoride is very toxic, and it corroded their laboratory equipment.

Making predictions

In 1789, French chemist Antoine-Laurent Lavoisier (1743–1794) predicted the existence of an unknown element. He made this claim after studying elements that had already been identified and realizing that one was missing. A few years later, French scientist André-Marie Ampère (1775–1836) realized that fluorspar held the key to this missing element, which he named fluorine. Ampère knew that fluorspar contained fluorine, but he did not know how to isolate the element from the mineral.

It took another hundred years for French chemist Henri Moissan (1852–1907) to isolate fluorine gas from fluorspar. He eventually succeeded using a process called electrolysis. Electrolysis is a way of splitting up compounds into their component elements using electricity.

Electrolysis involves running an electric current through a liquid that has the compound dissolved in it. When a compound dissolves in the liquid, it forms ions—atoms or groups of atoms with positive and negative charges. Ions with positive charges will move toward the negatively charged electrode, called the cathode. Ions with negative charges will move toward the positively charged electrode, called the anode.

A portrait of French chemist Henri Moissan taken in 1906—a year before his death. Moissan was the first person to study the properties of fluorine and its reactions with other elements.

Expensive equipment

The first chemists tried to isolate fluorine through the electrolysis of hydrofluoric acid (hydrogen fluoride dissolved in water), but they did not succeed. The hydrofluoric acid corroded their equipment, and most chemists did not realize that the chemicals they were using were dangerous and damaging to their health. Moissan solved all these problems by passing electricity through a mixture of hydrogen fluoride and potassium fluoride. Moissan plated his apparatus with platinum metal to stop the substances from corroding his equipment. The fluorine compounds reacted with the platinum, forming a skin of platinum fluoride

A reproduction of the electrolytic cell that French chemist Henri Moissan used to isolate the element fluorine in 1886. The apparatus is shown together with samples of the fluorine minerals cryolite (right) and fluorspar (left).

on the surface of the equipment, protecting it from further attack. The gases formed on the electrodes were kept separate to prevent an explosion. Using this costly apparatus, Moissan became the first person to see the pale yellow gas we know as fluorine. He was awarded the Nobel Prize for Chemistry in 1907 in honor of his pioneering work.

DID YOU KNOW?

MODERN PREPARATION

The modern preparation of fluorine is very similar to Moissan's technique, but it is done on a much larger scale. Potassium fluoride is mixed with hydrogen fluoride in a copper vessel, which is much cheaper than Moissan's platinum crucible. The electrodes are made of graphite—a form of carbon used in pencil leads. Graphite electrodes produce a huge current—up to twenty times stronger than that used to power a household appliance. The fluorine gas produced at the electrode is piped along steel tubes to copper containers. The fluorine gas is squeezed to a pressure twenty-seven times atmospheric pressure until it condenses into a liquid. Fluorine is then transported as a liquid in nickel canisters.

Chemistry and compounds

Fluorine gas is so reactive that it combines with almost all of the other elements. It even reacts with the most unreactive elements, such as gold and platinum. Fluorine also forms compounds with other members of its own group—the halogens—which is unusual.

A few elements do not form compounds with fluorine. These lighter elements are called inert or noble gases—helium, neon, and argon. All the inert gases have a full complement of eight electrons in their outer electron shells. Helium, neon, and argon hold their electrons very tightly, and even fluorine cannot react with them.

Why is fluorine so reactive?

Atoms react with each other to end up more stable with full outer electron shells. Atoms that need fewer electrons to become stable are more reactive. The key to fluorine's reactivity lies in the way it attracts electrons more strongly than any

A chemist holds a piece of steel wool in a stream of fluorine gas. The steel wool bursts into flames as soon as the gas touches it. This experiment demonstrates the highly reactive nature of fluorine.

DID YOU KNOW?

THE OXIDATION GAME

Oxidizing agents are substances that steal electrons from other substances. Fluorine is the most powerful oxidizing agent. A fluorine atom oxidizes other atoms by stealing electrons to fill up its outer electron shell. In this process the fluorine atom itself is said to be reduced—it gains electrons. The other atom is said to be oxidized—it loses electrons. The term *oxidation* was chosen because chemists first linked these reactions to the element oxygen.

When an atom is oxidized it is given a Roman number that counts for the number of electrons it has lost. This number is called the oxidation state. Fluorine is such a strong oxidizer that it produces the highest oxidation states in nature. In silver fluoride, for example, the silver has lost two electrons and has an oxidation state of +2. The compound is written silver (II) fluoride. The highest oxidation states are achieved when fluorine combines with other heavier halogens, forming compounds such as bromine (V) fluoride and iodine (VII) fluoride.

other element. Because fluorine and the other halogens need only one more electron to become stable, they are very reactive. Atoms that need at least two or three electrons to fill up the outer shell are less reactive. Metal atoms generally have only a few electrons in their outer shell. During a reaction these metal atoms are more likely to give up their electrons in the outer shell and shrink down to the full stable shell underneath.

ATOMS AT WORK

Hydrogen gas and fluorine gas react to form hydrogen fluoride gas. This reaction can be highly explosive. Hydrogen gas consists of a pair of hydrogen atoms bonded together. Fluorine gas consists of a pair of fluorine atoms bonded together.

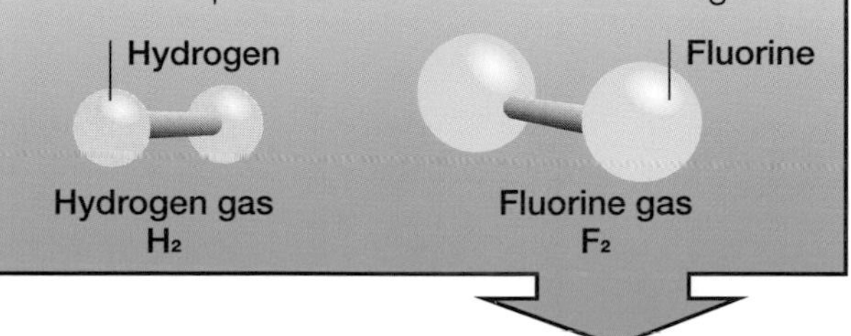

During this reaction, the bonds between the hydrogen molecules (H_2) and fluorine molecules (F_2) break apart. The hydrogen and fluorine atoms recombine to form two molecules of hydrogen fluoride (2HF).

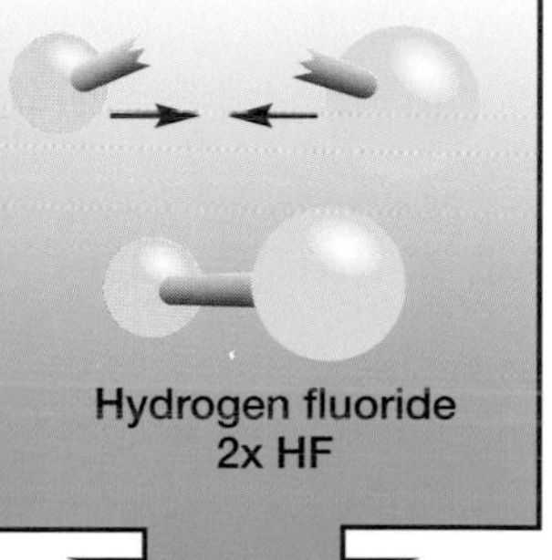

When the hydrogen fluoride molecules dissolve in water, they split up into hydrogen ions (H^+) and fluoride ions (F^-). As a solution, hydrogen fluoride is a very corrosive acid called hydrofluoric acid.

Hydrogen ions 2x H^+

Fluoride ions 2x F^-

The reactions that take place can be written like this:

$$H_2 + F_2 \rightarrow 2HF$$
$$2HF \rightarrow 2H^+ + 2F^-$$

The number of atoms of each element is the same on both sides of each equation, although the atoms have joined up in new combinations.

The high reactivity of gaseous fluorine means that it must be changed into a liquid and stored in pressurized nickel canisters.

Electronegativity

An element that attracts electrons is described as being electronegative. Fluorine is the most electronegative of all the halogens, because its outermost electron shell is closer to its nucleus. The nuclei of all atoms contain positively charged protons. The protons hold negatively charged electrons around the nucleus with a force similar to one between two magnets. The electrons are arranged in electron shells—one inside the other. Fluorine has only two electron shells, while chlorine has three shells, bromine has four, and iodine has five. The nucleus holds the electrons in the shells closest to it more strongly than those in shells farther away. So the force between the protons in the nucleus and the electrons in the outer shell is strongest in fluorine, then weaker in chlorine, with iodine having the weakest attractive force.

Stable substances

Atoms in most solid fluorine compounds exist as ions. Fluorine atoms steal electrons from less electronegative elements, and the resulting ions arrange themselves in a regular array called a crystal lattice. In a crystal of fluorspar (calcium fluoride; CaF_2), for example, each calcium ion is surrounded by eight fluoride ions and each fluoride by four calcium ions. Because of fluorine's electronegativity, fluorides such as fluorspar are very stable and unreactive substances. The only way to free fluorine from these stable compounds is to use electrolysis.

Charged particles

Some fluorine compounds do react with other substances. Some solid fluorine compounds dissolve in solvents, such as water, to form ions. Fluorine gas compounds are generally held together by covalent (shared electron) bonds, but they also form ions when they are dissolved in water. Fluoride ions are negatively charged because they have one extra electron in the outer shell. The other part of the broken-down fluorine compound forms a positively charged ion because it has lost electrons to the fluoride ion. Metal fluorides and hydrogen fluoride are most likely to form ions in solvents such as water, because they have only a few electrons in their outer shells and lose them fairly easily. When dissolved, the fluoride ions and metal or hydrogen ions are free to move around in the solvent, ready to react with anything added to the solution.

For example, hydrogen fluoride gas dissolves in water to form hydrofluoric acid—hydrogen ions and fluoride ions. If a piece of reactive metal, such as sodium, is dropped into hydrofluoric acid, hydrogen ions from the acid react with the sodium atoms. The hydrogen ions steal electrons from the sodium atoms to form sodium ions and hydrogen gas. This gas fizzes out of the water, producing a lot of heat. The fluoride ions are bystanders or "spectator" ions in this process.

ATOMS AT WORK

Fluorine gas consists of a pair of fluorine atoms bonded together. Sodium chloride solution consists of sodium ions and chloride ions dissolved in water. When fluorine gas passes through a sodium chloride solution, a chemical reaction takes place.

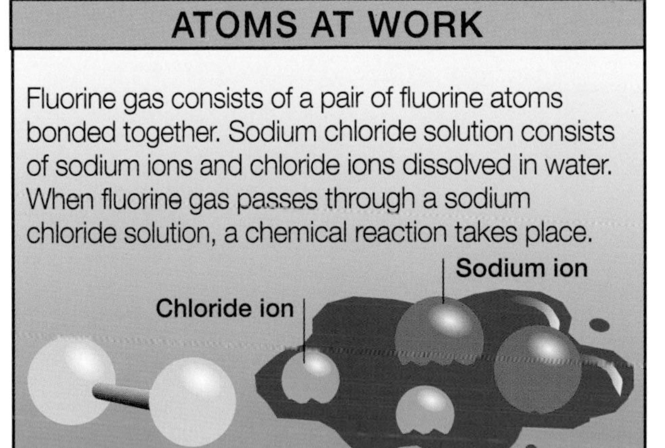

The bond between the fluorine atoms breaks apart. Each fluorine atom steals an electron from a chloride ion to form a fluoride ion. The chloride ion changes back into a chlorine atom. The two chlorine atoms then join up to form chlorine gas.

Electron e^- Electron e^-

This is called a displacement reaction, because the fluorine atoms have taken the place of the chlorine ions in the solution. A new compound called sodium fluoride can be made by boiling the water away.

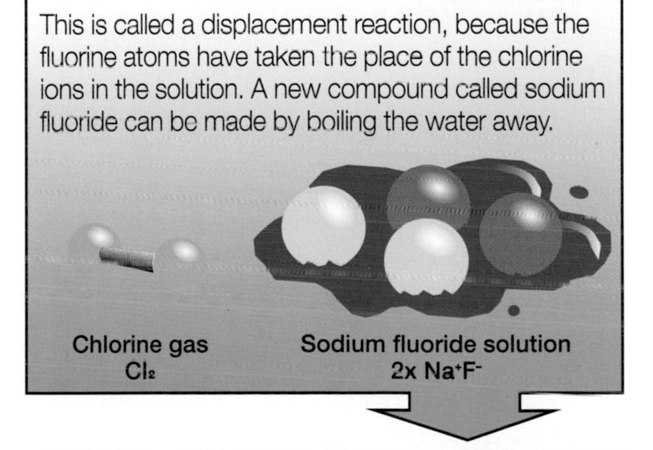

The reaction that takes place can be written like this:

$$2Na^+ + 2Cl^- + F_2 \rightarrow 2Na^+ + 2F^- + Cl_2$$

Fluorine and manufacturing

During electrolysis, pure aluminum collects at the bottom of a long bath called a reduction pot, which contains alumina dissolved in molten cryolite.

Fluorine's reactivity makes it a very important chemical in industry. For thousands of years, people used minerals containing fluorine compounds to purify metals without even knowing that fluorine existed. Even today, the largest use of a fluorine compound is in the refining of aluminum. This metal is extracted from an ore called bauxite, which contains a compound called aluminum hydroxide ($Al[OH]_3$). First, chemicals convert the bauxite into alumina (hydrated aluminum oxide; Al_2O_3). The bonds between the aluminum atoms and oxygen atoms are very strong. Electrolysis is the only way to break down the alumina into aluminum ions and oxide ions.

Electrolysis can only happen if the alumina is first dissolved in a bath of molten cryolite (sodium aluminum fluoride; Na_3AlF_6). The molten cryolite turns the alumina into a liquid so that the aluminum ions and oxide ions can move. When electricity passes through the bath, the aluminum ions move toward the lining of the bath, which forms the cathode of the electrolytic cell. There, each aluminum ion picks up an electron, forming an atom of pure aluminum. The oxide ions move toward the anode, lose electrons, and form oxygen gas.

In the intense heat of the bath, some of the cryolite breaks down to release more aluminum ions. Some of the aluminum extracted from the bath comes from the cryolite and not from the alumina. But the cryolite keeps reforming using the aluminum ions from the alumina. In this way, the cryolite does not run out as the process continues.

Silicon chips

One very important use of fluorine involves manufacturing silicon chips for computers and other products from the microelectronics industry. Fluorine compounds called perfluoros (PFCs) are used to etch patterns on silicon chips by removing parts of the thin layer of film that coats each chip. This process is called wafer patterning. PFCs are also used to remove the residue that clings to the chips. Once the residue builds up to a certain thickness, it peels off and creates particles that damage the chips.

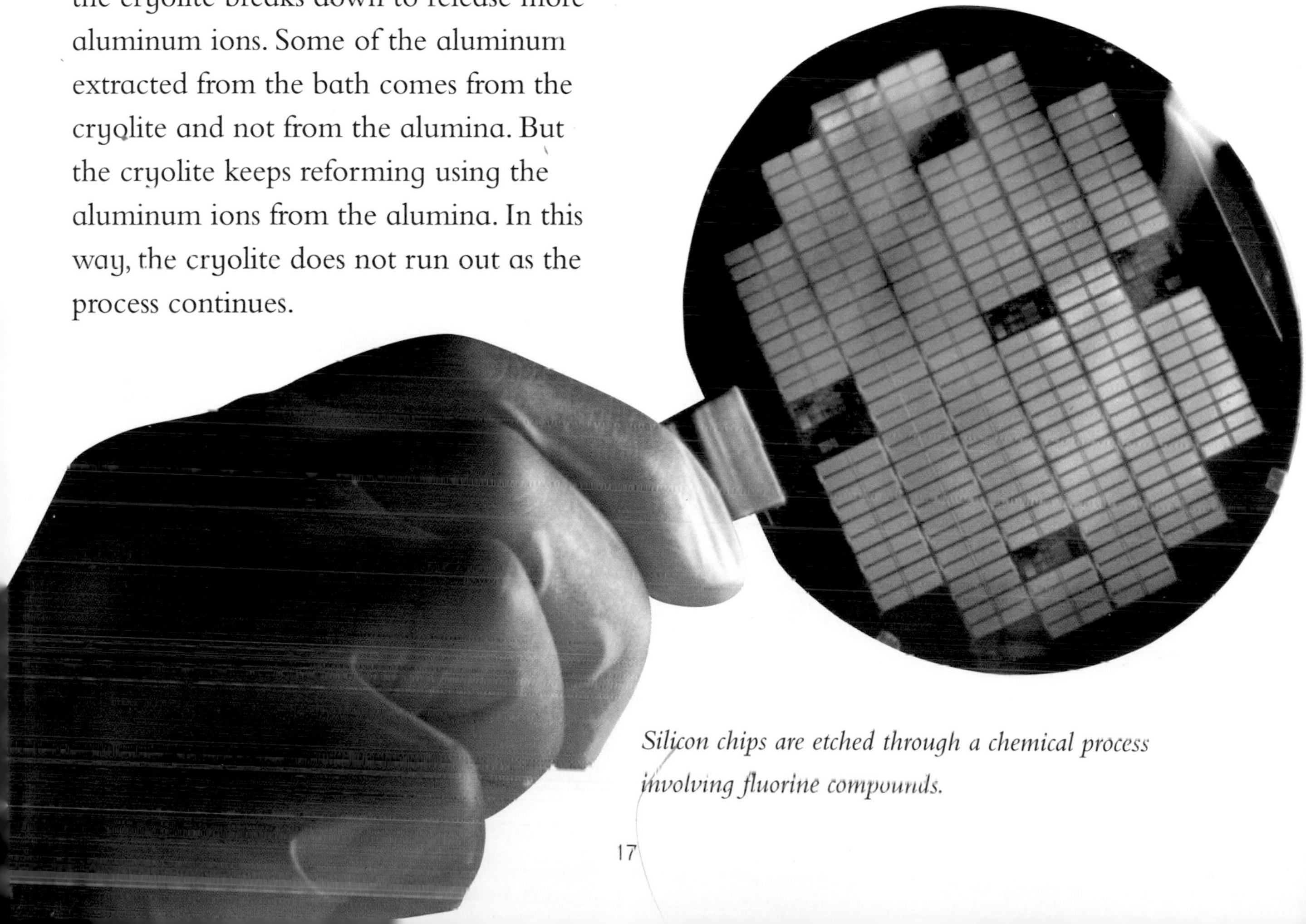

Silicon chips are etched through a chemical process involving fluorine compounds.

DID YOU KNOW?

CLEANROOMS

The tiny electric circuits etched on silicon chips have features that are as little as a few billionths of a meter across—just a tiny fraction of the width of a human hair. Any speck of dust or dirt in the air during the etching process may render the chip useless. For this reason, silicon chips are manufactured in cleanrooms. Cleanliness is the main consideration in these closed environments. Air-filtration equipment removes particles of dust in the air, and cleanroom workers pass through an air blast to remove any remaining particles off their special bodysuits before entering the cleanroom. Machinery is also specially designed to minimize contamination of the cleanroom.

Scientists have now developed new ways to etch patterns onto chips using fluorine gas lasers. The lasers emit light with a wavelength of 157 nanometers (one nanometer is one billionth of a meter). As a result, etchings on the chips can be very small. This new process solves two problems. The lasers do less damage to the environment than PFCs, and silicon chips can now have smaller circuitry.

Other uses

Fluorine compounds have many other uses. The first recorded use for hydrofluoric acid comes from a set of instructions for etching glass, which dates to around 1670. Hydrofluoric acid is still used in various forms of glass etching. It is used to mark divisions on thermometer tubes and to etch light bulbs and decorative patterns in glassware. This corrosive acid is also used in other forms of ceramic etching, such as pottery decoration. Other fluorine compounds are used to make tough coatings for ceramics and metals. Chlorine trifluoride (ClF_3) is a reactive compound that is used to cut through hot steel.

Many pesticides contain fluorine as an inactive agent. The fluorine in these pesticides does not actually kill the pests. It serves only to transport the active, pest-killing agent to the target. Other pesticides do contain fluorides as the active agents. One example is cryolite.

Hydrofluoric acid was used to etch the intricate patterning on this ornamental glassware.

Fluorine and fuel

Nuclear power plants make electricity using the heat produced by the fission of uranium fuel in a nuclear reactor. During fission, uranium atoms split up into two lighter elements. This process gives off a lot of radiation and heat energy. The radiation causes more

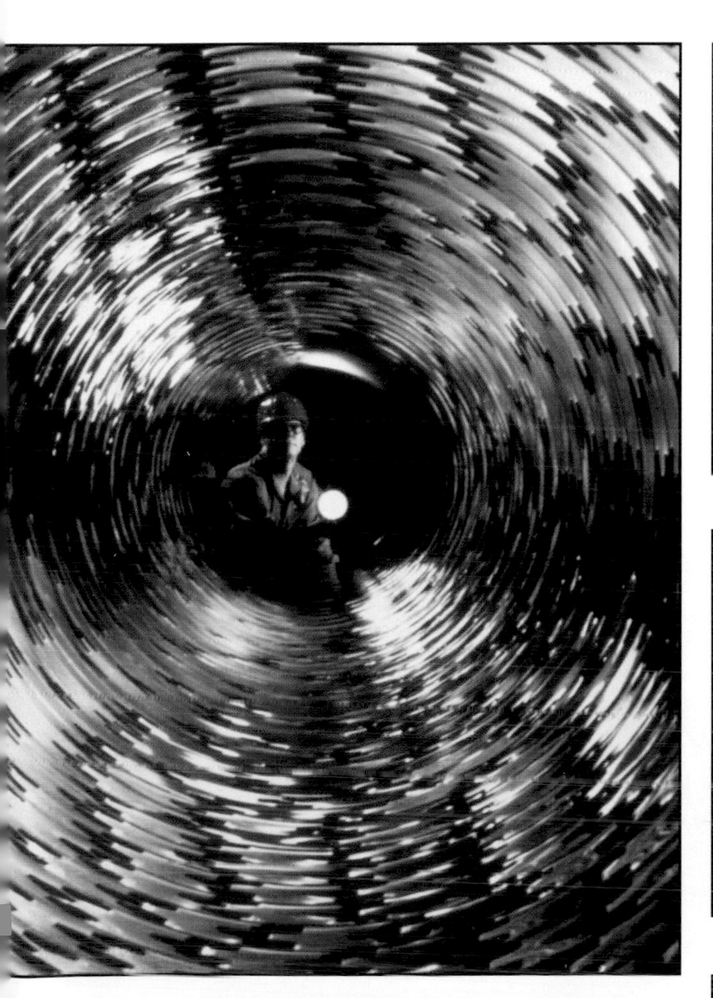

The inside of a uranium enrichment centrifuge, which is used to produce fuel suitable for use in a nuclear power plant.

ATOMS AT WORK

Uranium extraction is a complex process. The ore is dissolved in acid, heated, and then reacted with hydrogen to form uranium dioxide (UO_2). This reacts with hydrogen fluoride to form uranium tetrafluoride (UF_4). Uranium dioxide consists of one uranium bonded to two oxygen atoms. Hydrogen fluoride consists of one hydrogen bonded to one fluorine atom.

Oxygen
Uranium
Uranium dioxide
UO_2

Hydrogen
Fluorine
Hydrogen fluoride
HF

During the reaction the bonds between the hydrogen and fluorine atoms and the uranium and oxygen atoms break apart. Fluorine atoms replace the oxygen atoms in the uranium tetrafluoride molecule. Hydrogen and oxygen atoms are left over. They bond to form water molecules.

When fluorine gas is added to the uranium tetrafluoride, two more fluorine atoms join on to the uranium atom to form uranium hexafluoride gas (UF_6).

Uranium hexafluoride
UF_6

The reactions that take place can be written like this:

$$UO_2 + 4HF \rightarrow UF_4 + 2H_2O$$
$$UF_4 + F_2 \rightarrow UF_6$$

fission to occur, which produces more heat. The heat is then used to boil water, producing steam that drives a generator.

Uranium-235 is the only isotope that can be used as a nuclear fuel, but most uranium ore consists of uranium-238. So fluorine is used to separate uranium-235 from the uranium-238. The ore is first combined with fluorine to form uranium hexafluoride (UF_6) gas. Less than one percent of this compound contains uranium-235 atoms, so the gas is spun at high speed in an enrichment centrifuge. This forces the gas through a thin membrane filled with tiny holes. Uranium-235 moves through the membrane more quickly than uranium-238, so the gas beyond the membrane will be slightly richer in uranium-235. This "enriched" gas is collected and enriched again and again until it contains about 4 percent uranium-235. The gas is then converted for use as a fuel.

DID YOU KNOW?

ROCKET PROPULSION

Rockets are powered by the reaction of two substances—a propellant and an oxidizer. When they mix, they react violently. This causes a controlled explosion of heat and gases. The explosion is directed out of a nozzle at the bottom of the rocket, creating a thrust force that pushes the rocket into the air.

Since fluorine is the most powerful oxidizing agent in nature, it has been used in rocket propulsion in the past. When the National Aeronautics and Space Administration (NASA) was learning to build large rockets in the 1950s, they used liquid fluorine and liquid hydrogen to fuel many of their experimental models. Fluorine was the oxidizing agent and hydrogen was the propellant. The two liquids were pumped into a chamber at the base of the rocket. They reacted to produce hydrogen fluoride gas and a great deal of heat, which thrust the rocket up into the air. Pure fluorine is not used today, because it is hard to store, very expensive, and the hydrogen fluoride fumes produced by the rocket engine are very damaging to the environment.

NASA's Little Joe launch vehicle prepares for takeoff at the Langley Research Center in Wallops Island, Virginia. Liquid hydrogen and liquid fluorine were used to launch this experimental test vehicle.

Consuming small amounts of fluoride ions in drinking water may help promote the development of healthy teeth and bones. But many people fear that too much fluorine is bad for you.

Fluorine and health

Although fluorine gas is very toxic, most people think that compounds containing fluorine are too stable to harm the body. Fluoride ions are vital for good health. People need to have just a tiny amount of fluoride in their food and drink. Like many other essential minerals, too much fluorine can be dangerous.

Fluorine is required for the healthy development of teeth and bones. Most fluorine comes from drinking water, and many natural sources of water contain just the right amount of fluoride ions. However, in areas where the water does not have a natural supply of fluorine in it, studies have shown that children are more likely to have rickets—a disease that can deform bones—and people suffer from more tooth decay. To combat this problem, many health authorities now add small amounts of fluoride ions to drinking water in these areas. Adding fluoride ions

This picture shows a worker for the State Health Department adding sodium fluoride to the drinking water of residents of Newburgh, New York, in 1950.

to the water supply is called fluoridation. Most people think that fluoridation is good for our health.

Fluoride ions are not found only in the water supply. Dentists often treat children's teeth with fluoride ions to make sure they are healthy. Toothpastes also contain some fluoride ions in the form of tin fluoride and sodium fluorophosphates, so people can keep their own teeth strong without having to swallow too much fluorine.

Health warning

Not everyone is happy about adding fluorine to drinking water. They think that too many fluoride ions cause health problems. Consuming too much fluorine leads to a condition called fluorosis. The teeth of people with fluorosis are mottled

DID YOU KNOW?

ARTIFICIAL BLOOD

Physicians are currently looking for ways to use fluorine compounds to make artificial blood for patients undergoing surgery. At the moment, patients are given human blood donated by volunteers, but this must be kept refrigerated, and the blood used must be exactly the same type as the patient's blood. There is also a drastic shortage of blood donors.

To solve the problem, physicians have made organic (carbon-containing) liquids called perfluoros (PFCs), which contain many fluorine atoms. PFCs are stable compounds, and they are also very good at dissolving the oxygen people need to breathe. In experiments, animals can breathe normally while completely immersed in PFCs, and a mixture of these chemicals with water has been shown to work as well as blood in the same animals for short periods. Scientists are now working to insure that PFCs do not harm humans. If the tests are positive, people who undergo surgery may one day have fluorine compounds running in their veins instead of blood.

with patches of white and gray, although the teeth are otherwise healthy. Recent research has shown that fluoridation may be linked to brittle bones in the elderly, and fluoride ions may cause certain cancers. Too many fluoride ions may also reduce a person's iodine intake, which can affect intelligence and the functioning of the thyroid gland.

On several occasions, people in some areas of the United States have been allowed to vote whether they want their water to have fluoride ions in it. On most occasions, they have decided against adding fluoride ions.

Other risk factors

Over the past sixty years, there has been an enormous expansion in the use of fluorine compounds in agriculture and industry. People are exposed to airborne fluoride ions from aluminum smelting, coal-burning and nuclear power plants, glass etching, petroleum refining, plastic manufacturing, phosphate fertilizer production, silicon-chip manufacturing, and uranium enrichment facilities. All these uses may pose a risk to our health.

Fluorine pollution monitoring equipment at a vineyard near Perth in Western Australia. The biggest concern over fluorine pollution is the potential for fluoride ions to enter the food chain and reach dangerous levels in livestock and food crops.

Fluorocarbons

Fluorine is used to make a very useful group of organic compounds called fluorocarbons. Organic compounds consist of chains of carbon and hydrogen atoms. These chains are called hydrocarbons. The atoms of other elements, such as fluorine, can join with the hydrocarbon chains to make organic compounds with different properties. Organic compounds are very important because they are the building blocks of all living things. Other important sources of organic compounds include coal and oil. These substances are the remains of dead trees and other plants that lived on Earth millions of years ago.

The use of chlorofluorocarbons (CFCs) in aerosol cans is responsible for the hole in the ozone layer above Antarctica (the dark blue area in the picture below). CFCs are now banned in most countries, and the ozone layer is slowly recovering.

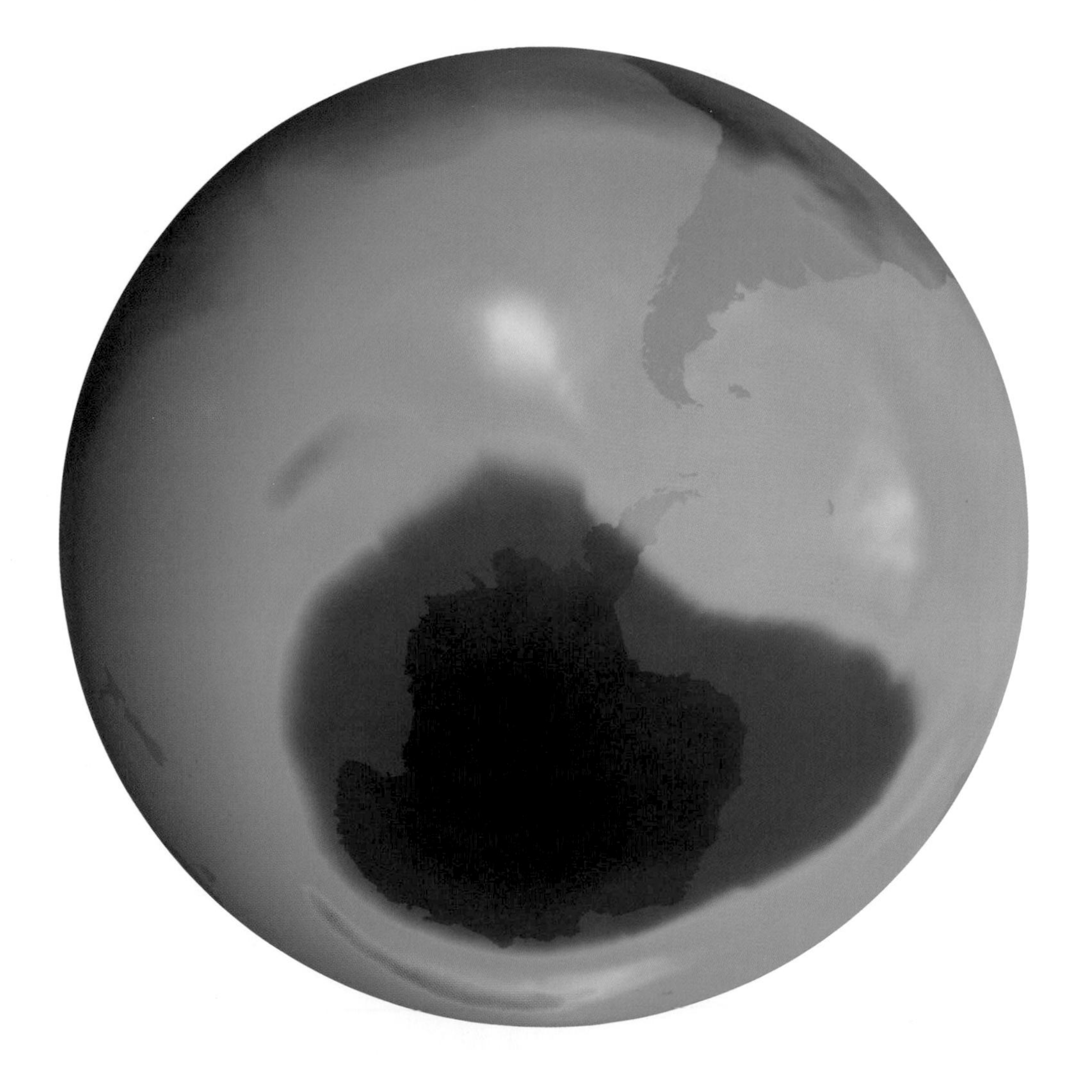

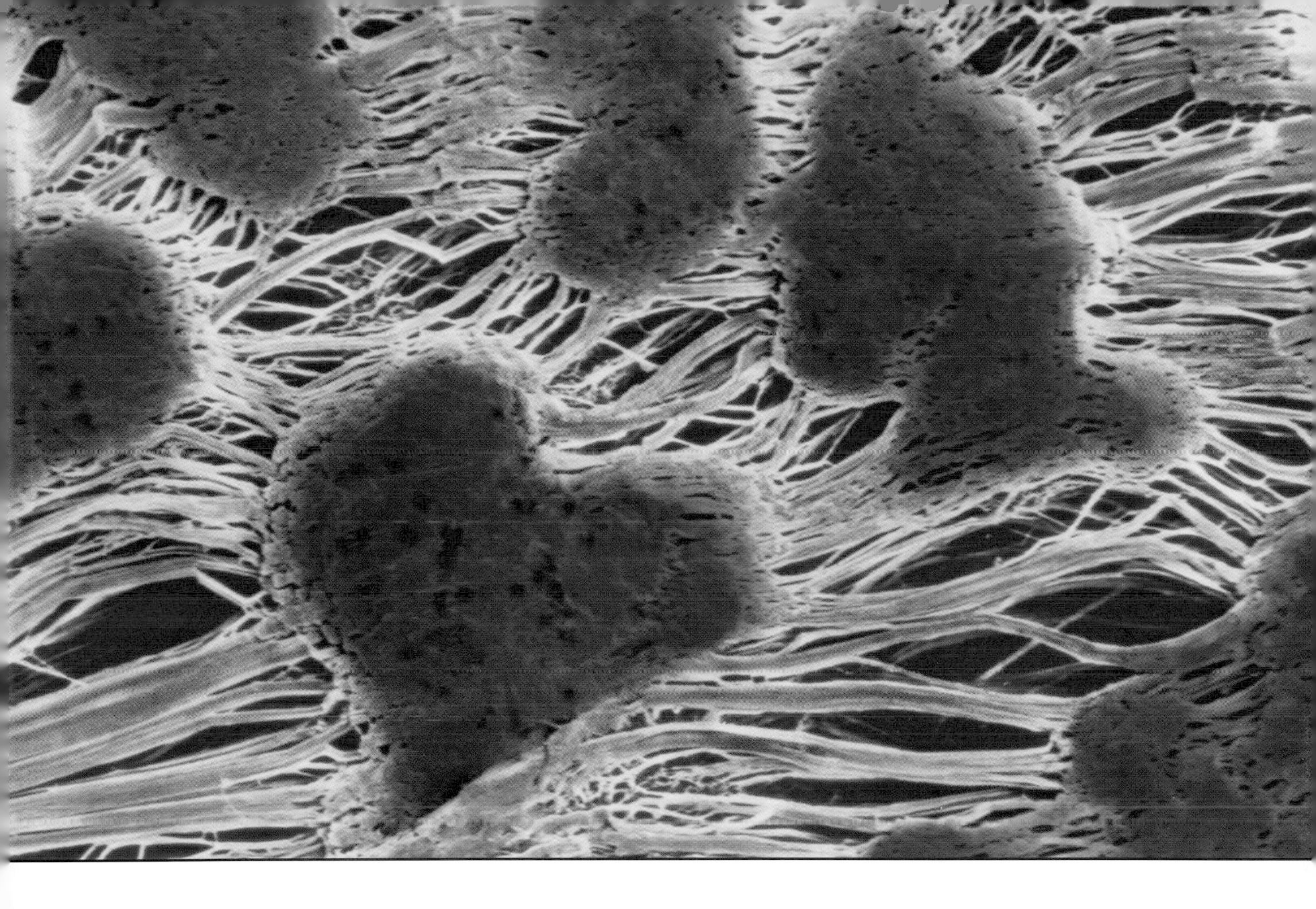

This colored electron micrograph shows a piece of stretched polytetrafluoroethylene (PTFE) tape—commonly called Teflon tape. Plumbers wrap Teflon tape around the threads on the end of a water pipe before screwing on the next pipe. The tape forms a watertight seal around the joint.

Stable compounds

Fluorocarbons are organic compounds where some or all of the hydrogen atoms in the hydrocarbon chain have been replaced by fluorine atoms. Since no other element is more reactive than fluorine, it is impossible to remove a fluorine atom from a fluorocarbon by chemical means. Indeed, fluorocarbons are some of the most stable organic compounds, and they are nontoxic. Fluorocarbons are often used when people need a liquid or gas that does not react with other things. For example, fluorocarbons called freons were once used as coolants in refrigerators and in air-conditioning systems.

Coolants and propellants

Freons belong to a group of organic compounds called chlorofluorocarbons (CFCs). As well as containing fluorine atoms, CFCs also have chlorine atoms joined to the hydrocarbon chain. CFCs are ideal for use as coolants in refrigerators because they boil and condense at just the right temperatures,

DID YOU KNOW?

TEFLON

Teflon is the tradename for a fluorocarbon called polytetrafluoroethylene (PTFE). This remarkable substance has many applications, ranging from nonstick cookware to waterproof insulation tape.

Teflon was discovered by accident in 1938, when a chemist stored tetrafluorethylene gas in a cylinder. After a few days, he could not get any of the gas to come out of the container, so he cut it in two to find out what was left inside. He found a waxy layer all around the inside of the cylinder where the tetrafluorethylene gas molecules had joined together to make a polymer.

The PTFE polymer was very unreactive, safe to use, and also amazingly slippery. Very little sticks to PTFE, and it also has a high melting point. It was not very long before someone thought to use Teflon as a nonstick coating on cookware. Nonstick pots and pans are easy to cook with because hot food does not burn and stick to the bottom. As a result, Teflon pans are much easier to clean.

and they do not corrode or damage the machine. CFCs were also once used as the propellants in aerosol spray cans. They were stored under pressure as a liquid inside the can, mixed with the bug killer, hairspray, or whatever needed to be sprayed. When someone pressed the nozzle, the liquid mixture shot out of the can and sprayed out as a mist of fine droplets into the air.

In the 1970s, environmental scientists found out that CFCs were damaging a protective layer in Earth's atmosphere called the ozone layer. Ozone gas is a special type of oxygen molecule with three oxygen atoms in it. The ozone layer protects life on Earth from the harmful ultraviolet light from the Sun. The scientists found that CFCs in the air were

Two chemists hold containers of tetrafluoroethane. This fluorocarbon has replaced chlorofluorocarbons as the propellant in aerosol cans. Tetrafluoroethane is much less harmful to the environment.

reacting with the ultraviolet light, releasing chlorine atoms into the ozone layer. In turn, these chlorine atoms were reacting with the ozone molecules, turning them into oxygen gas. Giant holes started to appear in the ozone layer. People living under these holes were at greater risk of skin cancers produced by ultraviolet light. Most countries stopped producing CFCs in 1996, and other organic oils are now used in their place.

Periodic table

Everything in the universe consists of combinations of substances called elements. Elements consist of tiny atoms, which are too small to see. Atoms are the building blocks of matter.

The character of an atom depends on how many even tinier particles called protons there are in its center, or nucleus. An element's atomic number is the same as the number of its protons.

Scientists have found around 110 different elements. About 90 elements occur naturally on Earth. The rest have been made in laboratories.

All the chemical elements are set out on a chart called the periodic table. This lists all the elements in order according to their atomic number.

The elements at the left of the table are metals. Those at the right are nonmetals. Between the metals and the nonmetals are the metalloids, which sometimes act like metals and sometimes like nonmetals.

- On the left of the table are the alkali metals. These elements have just one electron in their outer shells.
- On the right of the periodic table are the noble gases. These elements have full outer shells.
- Elements in the same group have the same number of electrons in their outer shells.
- Elements get more reactive as you go down a group.
- The number of electrons orbiting the nucleus increases down each group.
- The transition metals are in the middle of the table, between Groups II and III.

Group I	Group II	Transition metals						
1 H Hydrogen 1								
3 Li Lithium 7	4 Be Beryllium 9							
11 Na Sodium 23	12 Mg Magnesium 24							
19 K Potassium 39	20 Ca Calcium 40	21 Sc Scandium 45	22 Ti Titanium 48	23 V Vanadium 51	24 Cr Chromium 52	25 Mn Manganese 55	26 Fe Iron 56	27 Co Cobalt 59
37 Rb Rubidium 85	38 Sr Strontium 88	39 Y Yttrium 89	40 Zr Zirconium 91	41 Nb Niobium 93	42 Mo Molybdenum 96	43 Tc Technetium (98)	44 Ru Ruthenium 101	45 Rh Rhodium 103
55 Cs Cesium 133	56 Ba Barium 137	71 Lu Lutetium 175	72 Hf Hafnium 179	73 Ta Tantalum 181	74 W Tungsten 184	75 Re Rhenium 186	76 Os Osmium 190	77 Ir Iridium 192
87 Fr Francium 223	88 Ra Radium 226	103 Lr Lawrencium (260)	104 Unq Unnilquadium (261)	105 Unp Unnilpentium (262)	106 Unh Unnilhexium (263)	107 Uns Unnilseptium (?)	108 Uno Unniloctium (?)	109 Une Unnilenium (?)

Lanthanide elements	57 La Lanthanum 139	58 Ce Cerium 140	59 Pr Praseodymium 141	60 Nd Neodymium 144	61 Pm Promethium (145)
Actinide elements	89 Ac Actinium 227	90 Th Thorium 232	91 Pa Protactinium 231	92 U Uranium 238	93 Np Neptunium (237)

The horizontal rows of the table are called periods. As you go across a period, the atomic number increases by one from each element to the next. The vertical columns are called groups. Elements get heavier as you go down a group. All the elements in a group have the same number of electrons in their outer shells. This means they react in similar ways.

The transition metals fall between Groups II and III. Their electron shells fill up in an unusual way. The lanthanide elements and the actinide elements are set apart from the main table to make it easier to read. All the lanthanide elements and the actinide elements are quite rare.

Fluorine in the table

Fluorine is one of the halogens in Group VII of the periodic table. Like all the other halogens, fluorine has seven electrons in its outer electron shell. Fluorine is the most reactive of all the chemical elements. It reacts readily with all the other elements apart from the inert gases, forming a range of different compounds called fluorides.

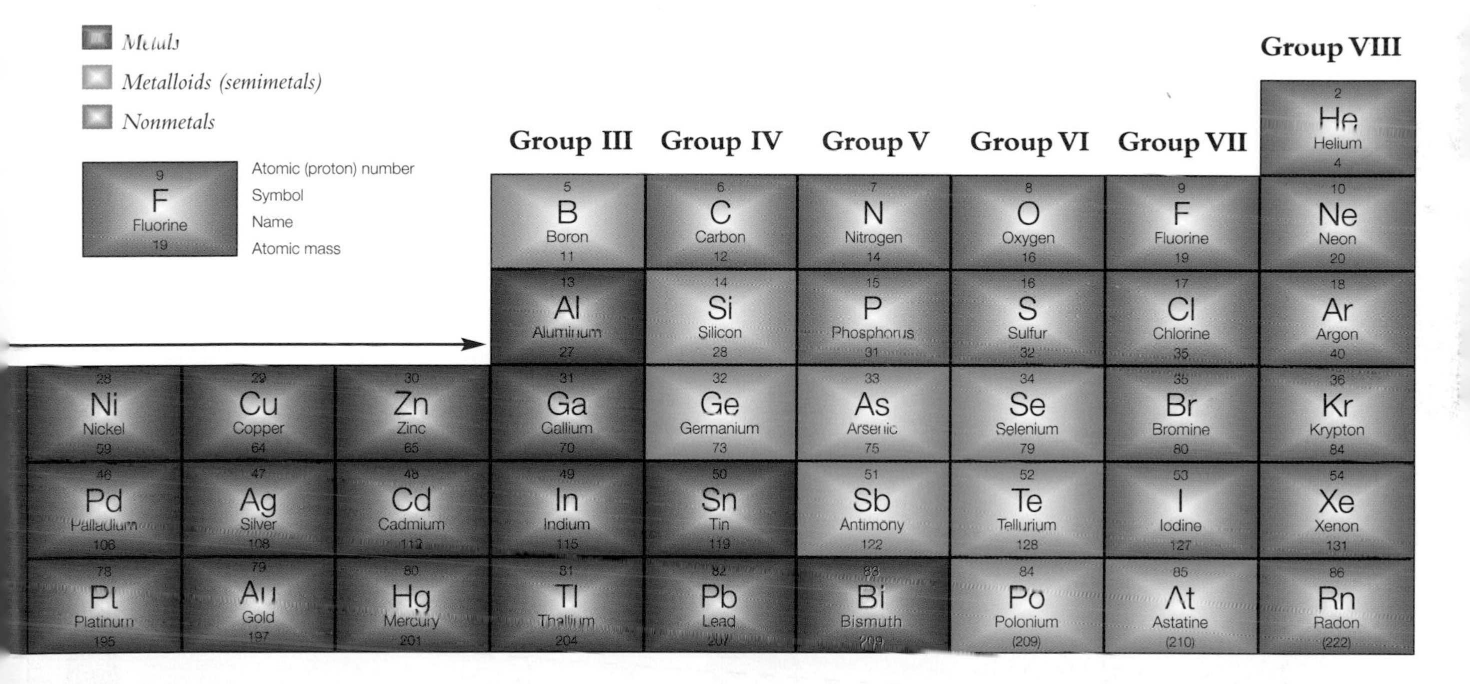

62 Sm Samarium 150	63 Eu Europium 152	64 Gd Gadolinium 157	65 Tb Terbium 159	66 Dy Dysprosium 163	67 Ho Holmium 165	68 Er Erbium 167	69 Tm Thulium 169	70 Yb Ytterbium 173
94 Pu Plutonium (244)	95 Am Americium (243)	96 Cm Curium (247)	97 Bk Berkelium (247)	98 Cf Californium (251)	99 Es Einsteinium (252)	100 Fm Fermium (257)	101 Md Mendelevium (258)	102 No Nobelium (259)

Chemical reactions

Chemical reactions are going on all the time—candles burn, nails rust, gasoline ignites in automobile engines, and food is digested. Some reactions involve just two substances, others many more. But whenever a reaction takes place, at least one substance is changed.

In a chemical reaction, the atoms do not change. A hydrogen atom remains a hydrogen atom; a fluorine atom remains a fluorine atom. But they join together in new combinations to form new molecules.

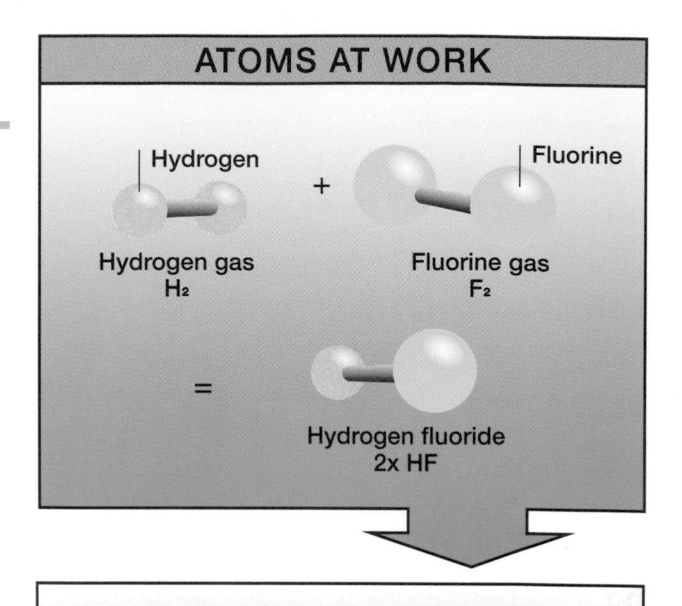

The reaction that takes place when fluorine gas reacts with hydrogen gas looks like this:

$$H_2 + F_2 \rightarrow 2HF$$

This tells us that one molecule of hydrogen gas reacts with one molecule of fluorine gas to give two molecules of hydrogen fluoride gas.

Writing an equation

Chemical reactions can be described by writing down the atoms and molecules before and the atoms and molecules after. Since the atoms stay the same, the number of atoms before will be the same as the number of atoms after. Chemists write the reaction as a chemical equation. Equations are a quick and easy way of showing what happens during a chemical reaction.

Making it balance

When the numbers of each atom on both sides of the equation are equal, the equation is balanced. If the numbers are not equal, something is wrong. So the chemist adjusts the number of atoms involved until the equation does balance.

Fluorine is never found as an element in nature. It is always combined with other elements in compounds such as this fluorspar crystal (calcium fluoride, CaF_2).

Glossary

atom: The smallest part of an element having all the properties of that element. Each atom is less than a millionth of an inch in diameter.
atomic mass: The number of protons and neutrons in an atom.
atomic number: The number of protons in an atom.
bond: The attraction between two atoms, or ions, that holds them together.
compound: A substance made of atoms of more than one element. The atoms are held together by chemical bonds.
crystal: A solid consisting of a repeating pattern of atoms, ions, or molecules.
electrode: A material through which an electrical current flows into, or out of, a liquid electrolyte.
electrolysis: The use of electricity to change a substance chemically.
electrolyte: A liquid that electricity can flow through.
electron: A tiny particle with a negative charge. Electrons are found inside atoms, where they move around the nucleus in layers called electron shells.
electronegativity: The tendency for an element to attract electrons.
element: A substance that is made from only one type of atom. Fluorine belongs to a group of elements called the halogens.
ion: A particle of an element similar to an atom but carrying an additional negative or positive electrical charge.
isotopes: Atoms of an element with the same number of protons and electrons but different numbers of neutrons.
metal: An element on the left-hand side of the periodic table.
mineral: A compound or element as it is found in its natural form on Earth.
molecule: A particle that contains atoms held together by chemical bonds.
neutron: A tiny particle with no electrical charge. Neutrons are found in the nucleus of almost every atom.
nucleus: The dense structure at the center of an atom. Protons and neutrons are found inside the nucleus of an atom.
organic compound: A compound that contains a chain of carbon atoms.
ozone layer: A layer of the atmosphere that protects Earth from ultraviolet rays.
periodic table: A chart containing all the chemical elements laid out in order of their atomic number.
proton: A tiny particle with a positive charge. Protons are found inside the nucleus of an atom.
radioactivity: The release of energy caused by particle changes in the nucleus.
solvent: A liquid that can dissolve or disperse other substances.
ultraviolet: Radiation similar to visible light, but invisible to the naked eye.

Index

alumina 16, 17
aluminum 4, 6, 7, 16, 17, 23, 29
Ampère, André-Marie 10
apatites 7

Bauer, Georg (Agricola) 9
blood (artificial) 22
bromine 4, 8, 13, 14, 29

chlorine 4, 6, 8, 14, 15, 18, 25, 27, 29
chlorofluorocarbons (CFCs) 24, 25–27
cleanrooms 18
cryolite 7, 11, 16, 17, 18
crystals 6, 7, 8, 14, 30

electrolysis 9, 10, 14, 16, 17
electrons 5, 6, 8, 12, 13, 14, 15, 17, 28, 29

fertilizers 7, 23
fluoridation 21–23
fluorocarbons 24, 25
fluorspar (fluorite) 6, 7, 9, 10,11, 14, 30
freons 25

halogens 4, 6, 8, 12, 13, 14, 29
hydrofluoric acid 8, 10, 13, 15, 18
hydrogen fluoride 9, 10, 11, 13, 15, 19, 20

iodine 4, 8, 13, 14, 23, 29
ions 10, 13, 14, 15, 16, 17, 21, 22, 23
iron 6, 9, 28
isotope 6, 8, 20

lasers 18
Lavoisier, Antoine-Laurent 9–10

Moissan, Henri 10–11

neutrons 5, 6
nuclear power 19–20, 23
nucleus 5, 6, 14, 28

oxygen 6, 7, 13, 16, 17, 19, 22, 26, 27, 29
ozone 24, 26–27

perfluoros (PFCs) 17, 22
pesticides 18
phosphates 7, 22, 23
platinum 10, 11, 12, 29
potassium fluoride 10, 11
protons 5, 6, 14, 28, 29

silicon chips 17, 18, 23
sodium 15, 17, 22, 28

teeth 4, 21–23
Teflon 25, 26
tetrafluoroethane 27
toothpaste 4, 22
topaz 7

ultraviolet light 26–27
uranium 19–20, 23, 28
uranium hexafluoride 19